D1511973

How Did I Get Here?

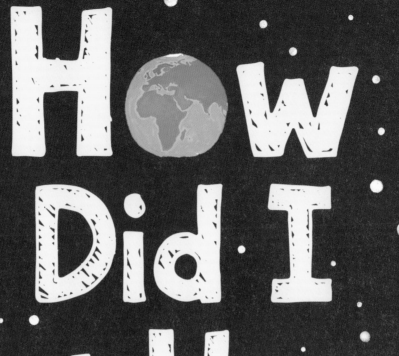

Your Story from the BIG BANG to Your BIRTHDAY

PHILIP BUNTING

(L)(B)

Little, Brown and Company

NEW YORK • BOSTON

Let's start at the beginning
(or at least what we think is the beginning).

Once upon a time, our entire universe
fit into a space smaller than an orange.

In those days, there wasn't much to see
around here. No light, no stars, no Earth.

That's heavy, man.

Until one day...

there was a really, really, really big...

Uh-oh.

bang!

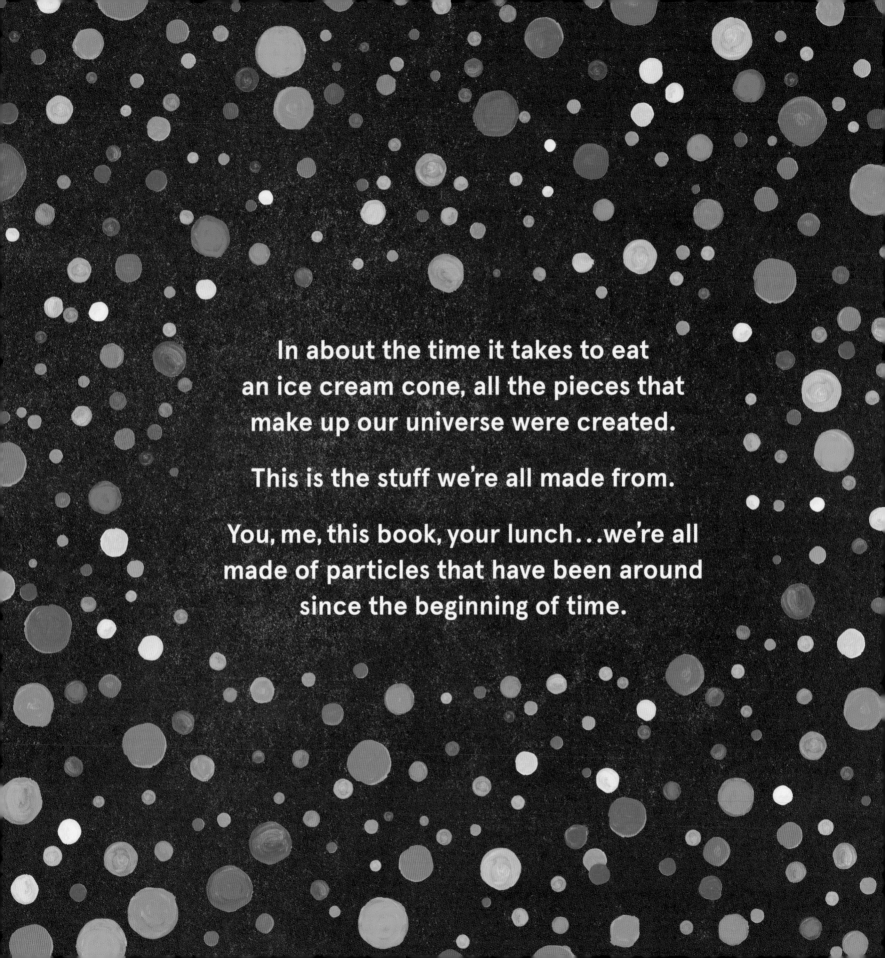

In about the time it takes to eat
an ice cream cone, all the pieces that
make up our universe were created.

This is the stuff we're all made from.

You, me, this book, your lunch…we're all
made of particles that have been around
since the beginning of time.

Hello.

Bonjour.

As they floated through the universe, some of these particles began to bump into one another. A few enjoyed each other's company so much that they decided to stick together.

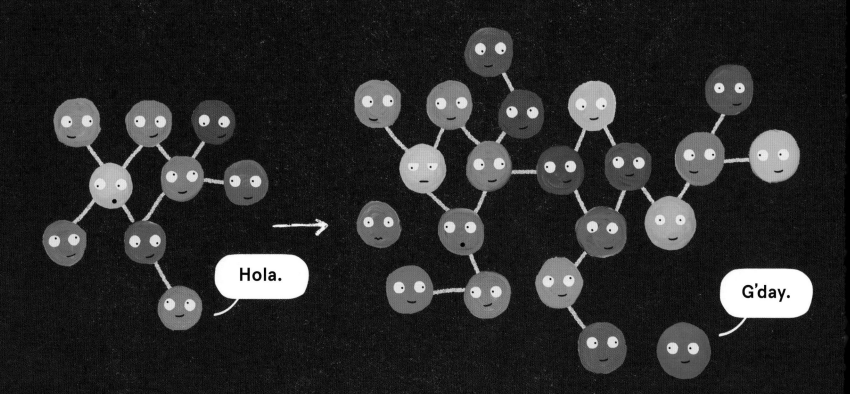

Hola.

G'day.

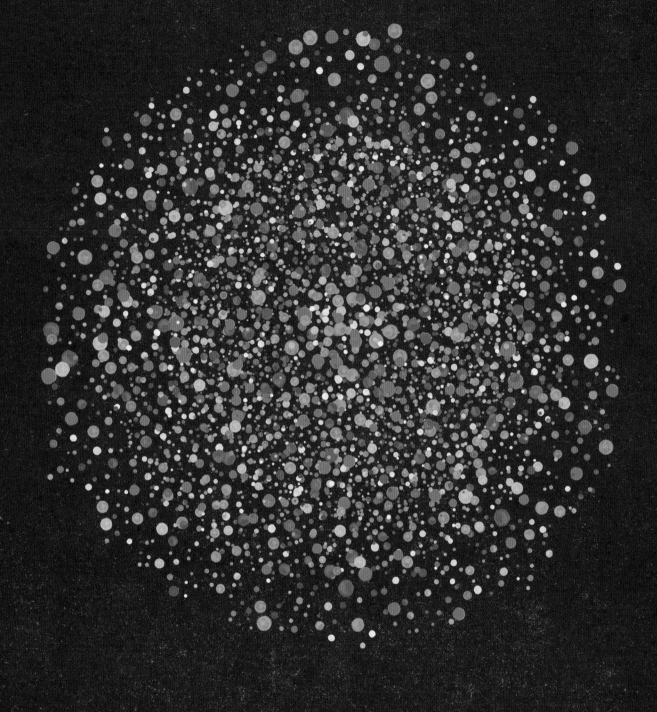

Eventually, so many particles stuck
together that they started to make things.
At first they formed ginormous
dust clouds.

These dust clouds attracted more and more particles.

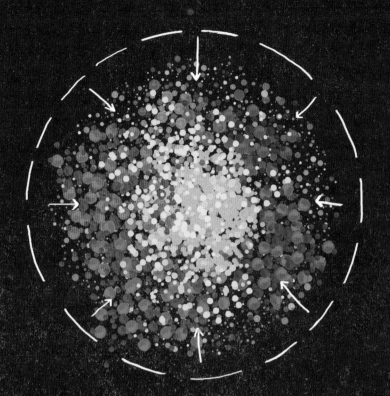

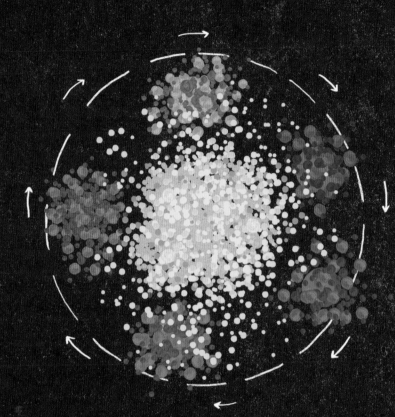

Over unfathomable lengths of time, they created suns.

And eventually, they created planets (which can't resist the pull of a good sun).

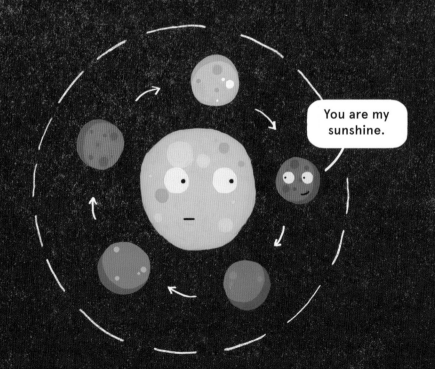

You are my sunshine.

Home
sweet
home

Our home planet is called Earth.

There we are, toward the middle of our
solar system. But Earth wasn't always
quite as homey as it is today.

At first, it was pretty warm around here.
But over time, Earth began to cool.

Aaah!

Aaah.

A long, long, long time ago

More recently

As our planet continued to cool, more and more
of those particles were drawn together.

More arrived from space, riding on meteors.

Some eventually became the land and water
on Earth's surface.

Then one day, many moons ago,
Earth was not too hot, and not too cold.
That warm water was just right for
the magic to happen.

Ta-da!

The first life-form was pretty simple. She couldn't see, hear, or wear a party hat. But she had one very special trick. She could make copies of herself.

Hello, beautiful.

This is where all of our stories start. All life on Earth came from this single-celled being (you can think of her as your great-great-great—times a trillion, zillion, squillion—granny).

You, me, the trees, worms, whales, and wolves... we're all related to this little lady.

All of life is one.

She might have been simple, but your great-granny^{gazillion} was no slouch.

Through generation upon generation, her children (your ancestors) slowly adapted to make the most of life in the warm waters of our early Earth.

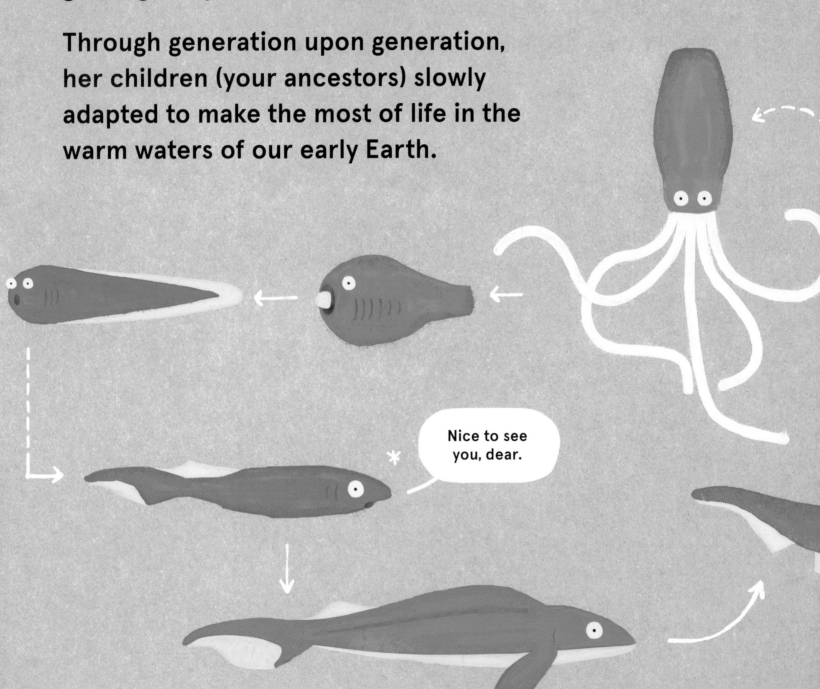

Nice to see you, dear.

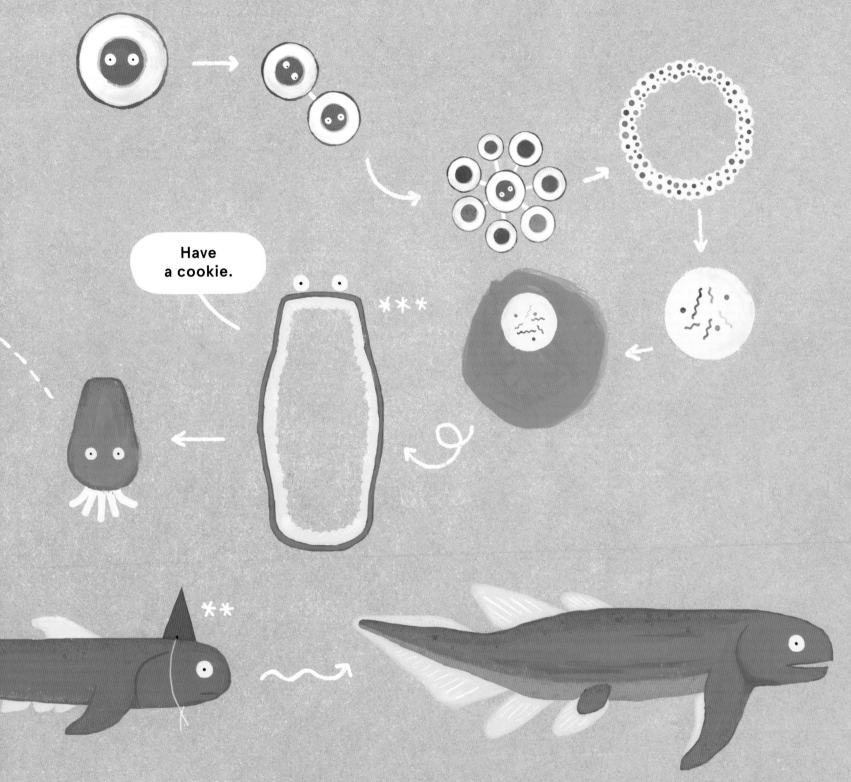

Have a cookie.

***

**

Very important small print:
* Our ancestors first developed eyes at about this stage in our journey. All peepers on creatures and creations before this point in the book have been gratuitously added for comic effect.
** Party hat also added for comic effect. Maybe it's her birthday.
*** This guy's name is Bob.

Then one fine day, one particularly unassuming but industrious little fish decided to see what she could see beyond the sea.

Her land-loving children would go on to become land animals. From dinosaurs to donkeys, yaks to you…we can all trace our family tree back to this adventurous amphibian.

Nothing to see here.

And of course, our evolutionary adventure didn't stop there. Over unimaginable lengths of time, our ancestors adapted to life on the land…

and then life
in the trees.

And from an ancestor we share
with chimps, we slowly evolved
into the species we are today.

The first humans lived in Africa, but our sense of curiosity and adventure soon took us to all corners of Earth.

Wherever you live on Earth, we are all descended from some very clever folks who lived right about here.

Except Antarctica. We left that to the penguins.

Thanks.

Eventually, we stopped wandering and began to settle down.

How me get here?

We learned how
to farm.

Fire!

We built
communities.

And then towns,
and cities.

Then one night, only a few years ago,
some of those particles—that were
once part of stars, and Earth, and
probably another life-form
or two—became you.

Oh my.

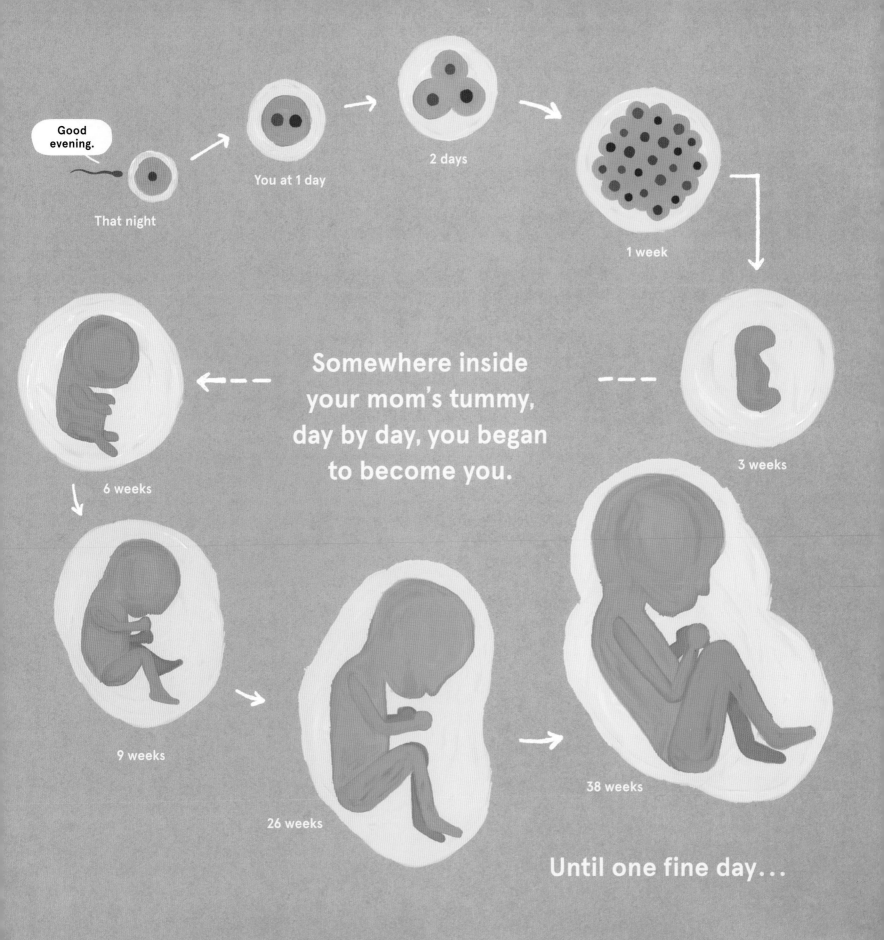

Good evening.

That night

You at 1 day

2 days

1 week

Somewhere inside your mom's tummy, day by day, you began to become you.

3 weeks

6 weeks

9 weeks

26 weeks

38 weeks

Until one fine day...

you made your grand entrance!

And that is how you got here.

You're very lucky to be here.

You are one of the newest additions to a family tree that goes all the way back to the very first life on Earth.

And just think, if any little thing had changed in the time after that first really, really, really big bang...

you might have turned out a little different.

FOR MY MUM, CAROL, WHO GOT ME HERE

Our universe is more unimaginably marvelous than we can ever suppose. We know so little. There's still so much more to discover about how we got here and where we're going. One thing we do know is that we humans have only one home (you're sitting on it). It's yours to take care of. If you do only one thing with this wild and beautiful life, make it your mission to leave Earth (and everyone who shares it with you) in better shape than when you got here.